INTELLIGENT HUMAN-MACHINE COLLABORATION

Summary of a Workshop

Ethan N. Chiang and Patricia S. Wrightson
Rapporteurs

Board on Global Science and Technology

Policy and Global Affairs

NATIONAL RESEARCH COUNCIL
OF THE NATIONAL ACADEMIES

THE NATIONAL ACADEMIES PRESS
Washington, D.C.
www.nap.edu

THE NATIONAL ACADEMIES PRESS 500 Fifth Street, NW Washington, DC 20001

NOTICE: The project that is the subject of this report was approved by the Governing Board of the National Research Council, whose members are drawn from the councils of the National Academy of Sciences, the National Academy of Engineering, and the Institute of Medicine. The members of the committee responsible for the report were chosen for their special competences and with regard for appropriate balance.

This study was supported by Contract No. HHM402-10-D-0036 between the National Academy of Sciences and the Department of Defense. Any opinions, findings, conclusions, or recommendations expressed in this publication are those of the authors and do not necessarily reflect the views of the organizations or agencies that provided support for the project.

International Standard Book Number-13: 978-0-309-26264-4
International Standard Book Number-10: 0-309-26264-X

Additional copies of this report are available from the National Academies Press, 500 Fifth Street, NW, Keck 360, Washington, DC 20001; (800) 624-6242 or (202) 334-3314; *http://www.nap.edu*.

Copyright 2012 by the National Academy of Sciences. All rights reserved.

Printed in the United States of America

THE NATIONAL ACADEMIES
Advisers to the Nation on Science, Engineering, and Medicine

The **National Academy of Sciences** is a private, nonprofit, self-perpetuating society of distinguished scholars engaged in scientific and engineering research, dedicated to the furtherance of science and technology and to their use for the general welfare. Upon the authority of the charter granted to it by the Congress in 1863, the Academy has a mandate that requires it to advise the federal government on scientific and technical matters. Dr. Ralph J. Cicerone is president of the National Academy of Sciences.

The **National Academy of Engineering** was established in 1964, under the charter of the National Academy of Sciences, as a parallel organization of outstanding engineers. It is autonomous in its administration and in the selection of its members, sharing with the National Academy of Sciences the responsibility for advising the federal government. The National Academy of Engineering also sponsors engineering programs aimed at meeting national needs, encourages education and research, and recognizes the superior achievements of engineers. Dr. Charles M. Vest is president of the National Academy of Engineering.

The **Institute of Medicine** was established in 1970 by the National Academy of Sciences to secure the services of eminent members of appropriate professions in the examination of policy matters pertaining to the health of the public. The Institute acts under the responsibility given to the National Academy of Sciences by its congressional charter to be an adviser to the federal government and, upon its own initiative, to identify issues of medical care, research, and education. Dr. Harvey V. Fineberg is president of the Institute of Medicine.

The **National Research Council** was organized by the National Academy of Sciences in 1916 to associate the broad community of science and technology with the Academy's purposes of furthering knowledge and advising the federal government. Functioning in accordance with general policies determined by the Academy, the Council has become the principal operating agency of both the National Academy of Sciences and the National Academy of Engineering in providing services to the government, the public, and the scientific and engineering communities. The Council is administered jointly by both Academies and the Institute of Medicine. Dr. Ralph J. Cicerone and Dr. Charles M. Vest are chair and vice chair, respectively, of the National Research Council.

www.national-academies.org

PLANNING COMMITTEE FOR THE WORKSHOP ON INTELLIGENT HUMAN-MACHINE COLLABORATION

JEFFREY M. BRADSHAW (Chair), Florida Institute for Human and Machine Cognition
DIANNE CHONG, The Boeing Company
GAL KAMINKA, Bar-Ilan University
GEERT-JAN KRUIJFF, Deutsches Forschungszentrum für Künstliche Intelligenz (DFKI)
BRIAN WILLIAMS, Massachusetts Institute of Technology

Principal Project Staff

WILLIAM O. BERRY, Director
ETHAN N. CHIANG, Program Officer
PATRICIA S. WRIGHTSON, Associate Director

BOARD ON GLOBAL SCIENCE AND TECHNOLOGY

RUTH DAVID, Analytic Services, Inc. (Chair)
HAMIDEH AFSARMANESH, University of Amsterdam
KATY BÖRNER, Indiana University Bloomington
JEFFREY BRADSHAW, Florida Institute for Human and Machine Cognition
DIANNE CHONG, The Boeing Company
JARED COHON, Carnegie Mellon University
ERIC HASELTINE, Haseltine Partners, LLC
JOHN HENNESSEY, Stanford University
NAN JOKERST, Duke University
PETER KOLCHINSKY, RA Capital Management, LLC
CHEN-CHING LIU, Washington State University
KIN MUN LYE, Singapore's Agency for Science, Technology and Research
BERNARD MEYERSON, IBM Corporation
KENNETH OYE, Massachusetts Institute of Technology
NEELA PATEL, Abbott Laboratories
DANIEL REED, Microsoft Corporation
DAVID REJESKI, Woodrow Wilson Center

Staff
WILLIAM O. BERRY, Director
ETHAN N. CHIANG, Program Officer
NEERAJ GORKHALY, Research Associate
PATRICIA S. WRIGHTSON, Associate Director

Preface

The Workshop on Intelligent Human-Machine Collaboration was organized by a planning committee whose role was limited to identification of topics and speakers. During its deliberations, the planning committee focused on topics that addressed the challenges and opportunities presented by intelligent collaboration between humans and machines. In acknowledging that interpretations of "intelligent" and "collaboration" vary among different scientific communities, the planning committee sought workshop participants from a range of science and engineering disciplines relevant to human-machine collaboration. Throughout the workshop, participants were not asked to arrive at consensus on any issue but, to explore human-machine collaboration issues from diverse disciplinary and cultural perspectives.

As such, the selected workshop topics and subsequent discussions were not intended to provide comprehensive coverage of all research efforts in the field of human-computer or human-robot interaction, but to glean insights into the research challenges and opportunities presented by intelligent human-machine collaboration in dynamic and unstructured environments.

The present summary was prepared by the rapporteurs as a factual summary of the presentations and discussions that took place at the workshop. Statements and opinions expressed are those of individual presenters and participants and are not necessarily endorsed or verified by the National Academies, and they should not be construed as reflecting any group consensus.

Acknowledgment of Reviewers

This report has been reviewed in draft form by individuals chosen for their diverse perspectives and technical expertise, in accordance with procedures approved by the National Academies' Report Review Committee. The purpose of this independent review is to provide candid and critical comments that will assist the institution in making its published report as sound as possible and to ensure that the report meets institutional standards for quality and objectivity. The review comments and draft manuscript remain confidential to protect the integrity of the process.

We wish to thank the following individuals for their review of this report: Henrik Christensen, Georgia Institute of Technology; Michael Goodrich, Brigham Young University; Paul Maglio, University of California; and Branko Sarh, The Boeing Company.

Although the reviewers listed above have provided many constructive comments and suggestions, they were not asked to endorse the content of the report, nor did they see the final draft before its release. Responsibility for the final content of this report rests entirely with the rapporteurs and the institution.

Contents

1	Introduction	1
2	Scenario Exercises	3
3	Human-Machine Teamwork Panels	11
4	Common Challenges and Breakthroughs	17
5	Global and Transnational Issues	23
6	Revisiting the Scenarios	25

APPENDIXES

A	Workshop Participants	31
B	Workshop Agenda	33
C	Presentation Abstracts	39

1

Introduction

On June 12-14, 2012, the Board on Global Science and Technology held an international, multidisciplinary workshop in Washington, D.C., to explore the challenges and advances in intelligent human-machine collaboration (IH-MC), particularly as it applies to unstructured environments. This workshop convened researchers from a range of science and engineering disciplines, including robotics, human-robot and human-machine interaction, software agents and multi-agent systems, cognitive sciences, and human-machine teamwork. Participants were drawn from research organizations in Australia, China, Germany, Israel, Italy, Japan, the Netherlands, the United Arab Emirates, the United Kingdom, and the United States.

On day one of the workshop, participants worked in small, interdisciplinary groups to determine how advances in IH-MC over the next two to three years could be applied to solving a variety of different real-world scenarios in dynamic unstructured environments, ranging from managing a natural disaster to improving small-lot agile manufacturing.

On day two, participants organized into small groups for a "deeper dive" exploration of four research topics that had arisen during the scenario discussions. Later in the afternoon, the full group discussed IH-MC in terms of common challenges, hoped-for breakthroughs, and the national, transnational, and global context in which this research occurs.

On day three, participants again organized into small groups to focus on longer term research deliverables. In addition, ten participants gave presentations on their research, with topics ranging from human-robot communication, to disaster response robots, to human-in-the-loop control of robot systems.

Throughout the workshop, participants were not asked to arrive at consensus on any issue but, rather, to identify challenges and opportunities from different disciplinary and cultural perspectives.

What Is Intelligent Human-Machine Collaboration?

Prior to the workshop, participants were asked to give their own definition of "intelligent human-machine collaboration" in a preparatory questionnaire. The following samples show the rich diversity of their responses:

... machines and humans combining each other's strengths and filling-in for their weaknesses and empowering each other's capabilities;
... joint and coordinated action by people and computationally based systems, in which each have some stake in the outcome or performance of the mission;
... humans AND machines jointly perform tasks that they would not be able to perform on their own;
... integration of AI into machines;
... humans and machines are able to mutually adapt their behavior, intentions, and communications;
... cooperation that mimics interactions between two humans;
... naturalness of the observed human-machine interaction;
... neither human nor machine treats the other as a disturbance to be minimized.
... machines being partners, and not a tool, for humans;
... technology that amplifies and extends human abilities to know, perceive, and collaborate;
... better overall performance of the mission, independently of how it was achieved;
... shared responsibility, authority, goals.

2

Scenario Exercises

Each group was allotted two hours to discuss their scenarios and then prepare a PowerPoint presentation of their findings and proposed solutions. Among the questions posed to each group were: What kind of progress can be demonstrated within two years? What could be done in years three to five? What issues are raised by including both software agents and robots as team members? In what circumstances is this system likely to fail when deployed in a real-world environment?

Scenario A: Preparing For and Managing a Major Disaster
Moderator: Michael Goodrich
Group Members: Michael Goodrich, Geert-Jan (GJ) Kruijff, Alex Morison, Daniele Nardi, Lin Padgham, Satoshi Tadokoro

> *Description: Mexico City's 18 million inhabitants live within 40 miles of Mount Popocatepetl, an active volcano that most recently erupted in 2000. Group A, a private enterprise dubbed "007 and Beyond," was given two years to develop a prototype for human-machine collaboration that would prepare for, respond to, and help with rebuilding following a major eruption. In this scenario, the group sought to address the life cycle of activities that constitute disaster management—from prediction through evacuation, to disaster mitigation and eventual reconstruction. The aim of this scenario was to consider how humans, robots, and software agents could co-manage a disaster and its aftermath.*

The group's moderator, Michael Goodrich, summarized the group's discussion. According to Goodrich, the group focused on what it determined to be the core human-machine collaboration challenge of managing a major disaster: a decision-support system that enables affected individuals, their families, individual responders, groups of responders, agencies, and centralized planners to give and receive needed, reliable (trustworthy), and timely information. The goal of such a system would be to enable individuals to make independent decisions in ways that ultimately support safety and survivability.

The group's system design included three networked components: a centralized information repository; role-specific clients (e.g., the Red Cross or emergency food relief programs) that both "push" data into the repository and "pull" information to facilitate decision making; and a scenario simulator that could explore and evaluate feasibility for various interventions. Robots that explore areas unreachable by (or unsafe for) humans following an eruption could also be system "clients." By exploring numerous scenarios in advance, the simulator could be used before a disaster to help design evacuation and responder protocols and after a disaster to help plan and manage search and rescue operations in real time. The group also designed the "iVolcano app" to facilitate interactions between the information repository and the humans, agents, and robots that use it. Thus, through iVolcano, people who are affected by the eruption could obtain critical information, such as where to find food, medical supplies, shelter, and water and where to charge their cell phones.

Goodrich also indicated that too much information can sometimes be as dangerous as too little in a major disaster—for example, if hundreds of people learned at approximately the same time where food was available, a stampede could ensue. The group saw two other potential problems: (1) many people would not willingly "push" information to a centralized data repository, either because of interagency tensions or possible concerns over privacy or trust, and (2) the system would need a method for differentiating the meanings of critical words. For example, *water* means "fire suppressor" to a fireman but something completely different to a nurse. Thus it would help if different word usages are mapped to a common ontology so that, in a time-critical situation, the person seeking information from the server isn't overwhelmed by irrelevant information.

The group suggested that by year two it would be possible to put into operation a "thin" server capable of integrating a lot of the information that already exists. Although this achievement would not be "earthshaking" (no pun intended, said Goodrich), it could be useful. The group thought that in years three to five it would be possible to deploy the interactive planning simulator tool. Several of the client programs would likely take longer to develop.

Scenario B: Small-Lot Agile Manufacturing
Moderator: Matthias Scheutz
Group members: Tal Oron-Gilad, Don Mottaz, Gopal Ramchurn, Matthias Scheutz, Lakmal Seneviratne, Brian Williams

> Description: George owns a small furniture company that builds one-of-a-kind furniture for its customers. As such piecemeal work negates economies of scale, he needs another way to generate profits. George retains the developers

of the Pengo9000 (the members of Group B), to create a coworker robot that will make it possible for him to triple his profits without adding manpower or major retooling costs.

The group's moderator, Matthias Scheutz, summarized the group's discussion. Scheutz initiated his discussion by commenting that the group found the exercise immensely challenging—so much so that solving this scenario required solving all of AI. Thus the group decided to separate the "spirit of the exercise" from a prototype that could potentially be available in a two-year time frame. Aspirationally, a collaborative robot would have natural language capabilities and would be able to learn and generalize from its lessons to real-world task completion. The robot would have sensing and perception capabilities that would, for example, enable it to distinguish between different kinds of wood, drill-bit requirements, and so on. It would have "common sense" knowledge, in addition to the domain knowledge necessary for understanding the commonsense meaning of words. For example, when someone is told to "go to the kitchen and turn the stove on," a human understands that he must go to the kitchen before he turns the stove on. A conventional robot might be expected to know that the word "and" refers to parallel, sequential, or temporal sequencing, but it would not have the intuitive capability to infer the correct meaning. The aspirational collaborative robot would be able to take directions from a combination of verbal and gestural cues. Finally, that robot would have perceptual and actuation capabilities that would enable it to find chairs in another room and then know which ones need to be drilled.

The group suggested that a robot could be developed within two years to fulfill certain tasks. It would have effectors for drilling and clamping and algorithms for planning and scheduling, as well as detecting and targeting objects. The robot would also understand simple instructions, such as "drill a hole into the chair," but it might not be able to do so repeatedly. In years three to five, the group posited that the robot would be able to pick up tools and learn how to use new tools. It would respond to more complex chained commands in combination with gestures and could detect errors. As a result, the robot would be a more active participant in the manufacturing process. Generally, though, the robot would still be very constrained in its capabilities, and after five years it would still not be a partner for the human furniture maker.

Scenario C: Hospital Service Robotics
Moderator: Candy Sidner
Group Members: Paul Maglio, Candy Sidner, Liz Sonenberg, Tom Wagner, Rong Xiong, Holly Yanco

Description: A large healthcare organization calculates the enormous sums spent in simply moving things—food, laundry, trash, wheelchairs, even pa-

tients—in a hospital environment. The aim of Group C was to design within two years a system for collaboration between hospital staff and patients, robots, and software agents.

The moderator, Candy Sidner, who spoke on behalf of the group, discussed how the group grappled with the complexity of designing an integrated system of humans, robots, and software agents that could significantly improve nonurgent hospital operations. In addition, many of the skills required to make such a system both useful and cost-effective—such as high-level language and locomotive skills, and high-level human-behavior-recognition skills—are still many years out from real-time operability.

The group observed that many factors contribute to these complexities. First, many of the potential system "users" would be people with no expertise in robotics or software agent systems. Second, it is often difficult to separate the urgent from the nonurgent in hospital settings. For example, hospital staff would want even the simplest delivery robot to communicate to the appropriate staff person that it had come across a patient who had fallen on the floor. To achieve this, the robot would have to know that finding someone on the floor was an anomaly, who was the right person to contact, and that its request for help had been received and acted upon. Completion of these tasks by robots is currently infeasible. Third, many issues of cultural and language diversity of both staff and patient populations (not to mention hospitals in urban versus rural settings or in advanced versus developing countries) exist. Thus, some members of the group speculated, robots and software agents would best be programmed to address a variety of cultural norms regarding gender differences, the notion of personal space, and concerns about safety, to name just a few.

According to Sidner, the group surmised that within two years it would be possible to deploy a "tug" robot capable of carrying things in a basket from Point A to Point B. They also suggested that it would be possible to deploy a virtual "my hospital friend" capable of engaging in simple language communications with humans and helping patients with tasks that are not medically critical, such as ordering meals or leaving the facility on patient discharge. An optimal system—parts of which could take 20 years or more to realize—would require breakthroughs in numerous subdisciplines related to human-machine collaboration. This system would include the following attributes: robots that can safely lift and carry patients; robots and agents that can engage in natural speech with humans; a networked system of robots and agents that can effectively communicate with each other and with relevant hospital staff; agents and robots that can successfully negotiate task priorities with humans (and with each other); agents and robots that are capable of prioritizing and carrying out requests from multiple operators; and robots and virtual agents that can interact appropriately with patients of varying ages, cognitive abilities, emotional states, and medical conditions.

During the Q&A session, a workshop participant asked what a robot would have to do to convince a human that its own immediate priorities are more important than the human's. Sidner answered that this presents a complex negotiation problem that is yet to be fully investigated by the science community. As an example, she pointed out that negotiation between humans and robots presumes advances in modeling wherein researchers understand the cognitive model that the robot has of itself and of the person with whom it is communicating. These advances, she explained, have yet to occur.

Scenario D: Virtual Team Training
Moderator: Mark Neerincx
Group Members: Michael Beetz, Jeffrey Bradshaw, Frank Dignum, Michael Freed, Yukie Nagai, Mark Neerincx

Description: A U.S. ship will soon be passing through the Straits of Hormuz, an area of high risk for terrorist attack. The ship's captain would like to have an on-board training system that will help crew members prepare for any possible encounter. Group D was charged with developing an agent-based system for virtual team training that mixes humans and software agents in ways that challenge and improve their team skills.

The moderator, Mark Neerincx, summarized the group's discussions. As he explained, the group sought to design a virtual training system that would learn along with the trainees so that (1) the system's feedback to the trainees would improve incrementally; (2) the lessons themselves would become more challenging as the trainees' capabilities grew; and (3) the system would provide team as well as individual feedback. In effect, a successful system would result in the coevolution of the virtual instructor, the students, and the software agents. The ideal system must have a degree of complexity, Neerincx suggested, to develop models of the trainees that the virtual instructor can use to provide useful feedback in an ongoing way, and to change the nature of the training as the trainees improve.

Within two years, the group suggested, it would be possible to establish a basic evolving framework if the following subtasks could be accomplished: (1) specifying ontologies that provide the basic foundation for how the virtual instructor will act and reason over time; (2) designing scenario-building tools; (3) developing templates for the use-cases; (4) developing a taxonomy of feedback rules; (5) developing both a task and user model; (6) creating feedback types that are appropriate to the templates; and (7) developing a small set of behavior detectors to trigger specific types of feedback. It would also be useful for the learners to provide feedback to each other. Such data would be fed back into the system (which is capable of pattern-finding) to improve future feedback and exercises. Neerincx suggested that adapting interactive language programs might help develop the feedback templates in the two-year time frame.

In truth, the group believed that just about all of these tasks would be difficult to accomplish within two years, or even in years three to five. Neerincx noted that work on serious games might help in this area, as well as research on emotion modeling.

Scenario E: The Personal Satellite Assistant[1]
Moderator: Terry Fong
Group Members: Terry Fong, Robert Hoffman, Andreas Hofmann, Dirk Schulz, Jean Scholtz, Manuela Veloso

> *Description:* The Enterprise *on television's* Star Trek *is a roomy place; in any actual spacecraft, however, space is at an extreme premium. This group's task is to develop a Portable Satellite Assistant (PSA)—flying spherical robots approximately three inches in diameter—capable of assessing hazards, monitoring conditions, and traveling within a spacecraft to places that an astronaut is too large to enter.*

The moderator, Terry Fong, presented the group's discussion. He explained that working on a spacecraft presents two unique challenges: First, an astronaut's time is precious and costly; they actually have little time to do "work," as most of life on board is consumed by "housekeeping" and such human functions as sleeping, exercising, and eating. Second, the ship's interior is extremely cramped, cluttered, and without a defined floor or ceiling. Thus it would be very helpful to have on board a small robot that could serve as an extra set of eyes and, perhaps, an extra brain. One prototype for such a PSA would be a spherical object about three inches in diameter. The PSA's key capabilities would be mobile sensing; monitoring standard procedures to detect anomalies and possibly alert the astronaut when things go wrong; supporting normal procedures such as providing astronauts with temporal cues (e.g., "The next step is this.") and spatial cues (e.g., by asking, "Did you look at this thing?" and then shining a laser pointer on something); and providing reference data to a crew member who is carrying out a piece of work.

The group designed the PSA to include the following technologies: cameras with zoom, 3-D, and color capabilities; sensors for reading temperature, barcodes, RFID, etc.; microphones; avionics for independent navigation; and wireless communication. With these technologies, PSAs could potentially assist astronauts by executing checklist procedures, an otherwise time-consuming task. PSAs would also have the capacity to view areas of the spacecraft that are out of

---

[1] Very limited PSA-like technologies have been tested in outer space. See, for example, http://psa.arc.nasa.gov/ and http://ssl.mit.edu/spheres/.

an astronaut's line of sight, model the environment and show changes or abnormalities over time, confirm that procedural models are being followed, and provide timing alerts that anticipate what is needed next.

Fong added that the PSA would have a model of the particular human it is assigned to and would be tasked to learn the preferences and work-related idiosyncrasies (e.g., left-handedness) of "its" astronaut. The PSA would also have to "know" and compensate should its astronaut becomes less alert over time. The PSA would also have to be sufficiently resilient to adapt to a revised plan if its astronaut changes the sequence of a task for good reason. To achieve this, humans and their PSAs would undertake joint training prior to their mission.

The advantage of having such an assistant is that PSAs do not criticize or take offense. The disadvantages follow from the advantages: robots are incapable of exhibiting human behavior and the "uncanny valley" problem is likely to arise. The group speculated that accomplishing fixed tasks and mobility were achievable within two years. The PSA's ability to change its models of the environment, task at hand, and so on and to observe and engage in unanticipated tasks could exist by years three to five.

Participants were asked prior to the meeting to give examples of successful Intelligent human-machine collaboration that are currently in use. The most popular responses were:

- Robotic surgery
- Google Search/search engines
- Siri
- Production systems where humans and robots work together (e.g., Kiva Warehouse Robotics)
- Flight management and navigation systems on commercial aircraft
- Intelligent vehicles (e.g., Google's unmanned vehicle)
- I have seen no successful examples

3

Human-Machine Teamwork Panels

The workshop participants began the second day's discussions with four panels that sought to plunge deeper into some of the issues that arose during the five scenario presentations. The topics varied from rethinking the user-vendor relationship in robotics procurement, to enhancing the planning capabilities of agents and robots, to the deep-level meaning of communication, to the potential for real collaboration between humans and robots. The panels addressed research challenges and, in some cases, suggested possible approaches.

Panel One: Design, Evaluation, and Training
Moderator: Robert Hoffman
Group Members: Michael Freed, Robert Hoffman, Don Mottaz, Mark Neerincx, Jean Scholtz

The panel moderator, Robert Hoffman, provided the panel's approach to the design-build-test-deployment process of human-machine systems. They found problems with the process at every stage. Users cannot describe what they want because they don't know what is possible. They also don't speak the same substantive language as the engineers who will build the systems, thus practically ensuring a mismatch between what the user wants and what the engineer will build. The human models that are used to construct human-machine systems are usually too crude; at the opposite end of the spectrum, such cognitive modeling architectures as Soar and ACT-R, while having many uses, may be too complex. Hoffman spoke of the paradox underlying the construction of human-machine systems: Although these systems would be better overall if they were based on more complex cognitive models, as the models become more complex, they also become more "brittle," thus changing the original requirements of the system. Next, system components are often built in isolation from each other, with the result that they don't fit with the overall workflow. Finally, user training of the system is too little and too late, and it tends to focus more on the designer's theory than the user's needs.

To fix these problems, the panel offered a wish list of changes to current practice. The process should: (1) base models on knowledge and meaning and not just on data; (2) include hypotheses in cognitive models to make them less rigid and more adaptive; (3) create the role of "modeler" who can bridge the worlds of the user and engineer-builder; and (4) colocate testbeds with deployed systems so that user involvement can be rich from the start. Moreover, (5) engineers should not only train but also mentor the users of the system so as to maximize their usefulness. In addition, product deployment should not be the end of the relationship between the user and the vendor but, rather, the beginning of a second stage of empirical study by the vendor to deal with the unintended consequences of the system (both positive and negative) once it is in place. This second stage will improve the usability of the system at that particular site while offering lessons to the vendor for the next generation of the system.

During the Q&A portion of this panel discussion, Lin Padgham suggested that a looser funding model that focuses on the end product as opposed to item-by-item accounting could result in a cocreative process that more accurately reflects the vendor's capabilities and the user's needs.

Panel Two: Intent Recognition, Execution Monitoring, and Planning

Moderator: Andreas Hofmann
Group Members: Michael Beetz, Tal Oron-Gilad, Andreas Hofmann, Paul Maglio, Dirk Shulz, Lakmal Seneviratne, Liz Sonenberg, Satoshi Tadokoro

The moderator, Andreas Hofmann, spoke on behalf of the panel. As he explained, the panel focused on the challenges of intent recognition, execution monitoring, and planning that are associated with the sense-deliberate-act loop (also known as the robotics paradigm). He explained that most sensor-based data is "noisy" and requires filtering for quick and correct evaluation. The panel suggested that more sophisticated algorithms based on plan context might be able to filter out the "noise" related to visual and tactile sensors. This notion led the panel to consider the planning phase of the robotics paradigm: How would the agent(s), robot(s), or mixed teams assess the success of the plan itself?

Hofmann noted that it is unrealistic to define the successful outcome of a plan in terms of specific assumptions going in. A more realistic strategy would be to continually evaluate the plan's success. Here the panel suggested that execution should include an evaluation capability that can give a probabilistic estimate of the plan's success. If the estimate goes below a certain threshold, a human operator would be called in to re-plan or somehow alter the original plan. The challenge, according to Hofmann, is to do this sooner rather than later in the course of the plan's execution. Another challenge is that it may be difficult to design appropriate predictors for estimating a plan's success because humans actually use a variety of methods to perform tasks. Thus it may be difficult to assess when execution has dipped below the expected threshold because there are, indeed, many potentially acceptable thresholds.

Next the panel turned to a basic problem of the planning phase of the robotics paradigm: in the real world, environments are uncertain and dynamic; moreover, sometimes, plans are simply infeasible. Because data keep changing, planning is computationally intensive. The challenge is for the planning phase to happen quickly enough to keep the loop robust. The panel speculated that incremental planning algorithms could address the changing data challenges. Hofmann also suggested collaborative plan diagnosis as a promising area of research. This method views plan failure as a diagnostic problem—algorithms look for conflicts that need to be resolved or constraints that need to be removed to make the plan feasible. Some members of the panel also suggested that planning domains could be made more realistic if they were defined by the robot's action capabilities.

Hofmann concluded his discussion with a set of questions about the mental modeling that constitutes the foundation of intent recognition, execution monitoring, and planning. What is the right level of abstraction—quantitative, qualitative, or hybrid models? How should shared plans be represented? How should agent resource capabilities be represented? How should human resource capabilities be modeled? How should the human psychological or operational safety model be represented? What are the best estimation model learning algorithms that support estimation and control?

Panel Three: Communication
Moderator: GJ Kruijff
Group Members: Frank Dignum, GJ Kruijff, Yukie Nagai, Daniele Nardi, Lin Padgham, Matthias Scheutz, Candy Sidner

GJ Kruijff, the moderator, provided a summary of the panel's discussions. Kruijff indicated that the panel addressed fundamental problems associated with communication—not simply the sharing of words and gestures but the depth of meaning that words and gestures represent. The panel's goal was not to solve these problems as much as to describe them. Every dimension of communication, Kruijff noted, is composed of multiple sub-dimensions that affect the communication process. For example, what are the tasks in which communication occurs: single events or repeated ones? Structured or unstructured? Well or poorly understood? How many actors are communicating? What kind of knowledge is necessary for communication: Domain specific? Common sense? What kind of communication is going to take place: Face-to-face or side by side?

In the realm of human-robot interaction, Kruijff remarked, there are two functions that describe collaboration: teamwork and taskwork. Teamwork in this context refers to humans and robots coordinating their behavior to accomplish a task. Taskwork refers to the "doing" of the task itself.

Even before the task itself is undertaken, the team must communicate how to coordinate the team's behavior: negotiating who does what, who is responsible for what, who is expected to succeed at a particular task, and so on. Plans may need to be adjusted, because things in the environment have changed. A robot needs to understand all these different aspects of the team's coordination as well as what it means to progress for itself and others in carrying out these activities. The robot also needs to be able to identify when it or others need help.

Within the context of carrying out a task, communication is used to build up shared beliefs, or "common ground," among the actors so that everyone on the team is at the same level of understanding. Traditional approaches to modeling and various other AI issues assume an objective model that everyone can map into. But from the perspective of communication, this is not the case. All the members of the team perceive and act subjectively. They have their own experiences and their own understanding of the world. This is particularly true for robots versus humans in the team context, Kruijff remarked.

How, then, is it possible to align all team members, given that people and robots perceive quite differently? Kruijff speculated on behalf of the panel that the problem of communication is how to fit everything together: simple communication, the social dimensions, collaboration in terms of planning and execution, and motivations and expectations. Scheutz ended the panel's presentation by observing that sharing the deep representation of meaning is difficult enough between one human and one robot; it will take considerable research to be able to achieve this at the level of multi-member teams.

Panel Four: Collaboration
Group Members: Terry Fong, Mike Goodrich, Alex Morison, Gopal Ramchurn, Manuela Veloso, Tom Wagner, Rong Xiong

In contrast to the other panels that chose a moderator to speak for the entire group, each member of the group discussed aspects of collaboration of interest to him or her. Alex Morison discussed how collaboration involves reciprocity; team members cannot achieve their own goals without helping others. This means that each team member gives up some of his own goals in order to help others and to accomplish the overall mission. Yet in the world of human-robot interaction, Morison noted, reciprocity is not necessarily standard operating procedure. If a pilot sees a UAV, for example, he has orders to get out of the way because he cannot be sure what the UAV is going to do. Thus collaboration is still a work in progress for human-robot teams.

For Fong, collaboration should be seen as a spectrum from loosely coupled—even independent—coactivity to tightly coupled interaction. He suggested that a team can be productive as long as it coordinates what it does. Organization, which he defines here as the allocation of tasks, is central to successful coordination.

According to Rong Xiong, evaluation is an integral component of collaboration. Robots need a basis for self-evaluation that is derived from shared information. Similarly, when a human needs help, he needs to know what the robot can do and how to ask for it.

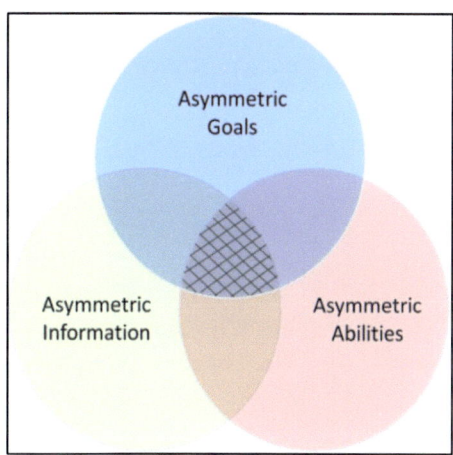

Goodrich would like to see collaboration research over the next ten years take place in the central area of overlapping circles (hatch marks) of the Venn diagram (shown left). Multiple human-robot teams have disparate or asymmetric goals, information, and abilities. Understanding collaboration will involve accounting for and aligning these asymmetries. Goodrich's understanding of collaboration is relevant to comments made by Jeff Bradshaw during the previous breakout session, in which he described seven myths related to autonomous systems. Taken together, these myths suggest that autonomy is more multidimensional, complex, and collaborative than is often viewed in the literature.

Manuela Veloso suggested that the robot's planning algorithms—such as Partially Observable Markov Decision Processes (POMDP), which enable robots to plan paths under partially observable conditions—would benefit from including models of the human that the robot may encounter in its environment. This will help the robot infer human intentions. In contrast to Goodrich's approach to collaboration, Veloso questioned whether complexity is really necessary for collaboration. Does the robot need to know why the human needs it to do a particular task, such as "go to the door"? In her view, it would be a great contribution just to be able to coordinate on minimal knowledge of intentions or needs.

Tom Wagner defined coordination as the process of managing interdependencies between tasks or plans and suggested taking Veloso's POMDP approach to the next step: to make an explicit representation of interdependence that would enable a robot to divert a human's attention to help it. For example, a robot waiting at an elevator would, instead of waiting opportunistically for the elevator door to open, ask a person walking by to press the button for it.

Gopal Ramchurn suggested that research is still to be done to find the balance between interaction design and mechanism design so that rules of engagement between and among humans and robots take incentives of team members into account.

4

Common Challenges and Breakthroughs

After the scenario and panel discussions, workshop participants discussed common challenges in IH-MC, as well as breakthroughs to transform the ways in which humans and machines will collaborate in the future. Some of these breakthroughs, discussed below, were also touched upon during the scenario exercises.

Rethinking Roles for Humans and Machines

For some participants, achieving the desired breakthroughs begins with a reevaluation of the role humans play in IH-MC. According to Kruijff, this begins with recognizing that humans have a central role. Rather than focusing on the human as a part of the problem or only as a partial solution, he proposed that collaboration should be designed with and around humans and take into consideration the broader sociotechnological context.

From the machine side, Sidner observed that current robots have limited sensing, manipulation, and communication capabilities. However, as machine capabilities improve, machines will be able to play a larger and more complex role in IH-MC. For some observers, this could result not only in new roles for machines in human-machine collaboration, but also a shift in human-machine team dynamics. Beyond humans leading teams of machines, Hoffman hypothesized that one day computational devices could serve as mentors or trainers for human (and/or human-robot) teams. If that were to happen, he asked, how might humans be trained to work with machine partners or even machine mentors? Understanding these questions, according to Liz Sonenberg, will benefit from a better understanding of what makes a human a good member/leader of a human-machine or all-machine team.

Jeff Bradshaw suggested that intelligent human-machine systems may also have a role in virtual and real-world training activities. For example, by designing experiments that require individuals to adapt in a changing environment, both on the field and in the laboratory, researchers could study individual and team dynamics. There was a debate over how closely human-machine training could replicate human team training (given the elaborate physical, cognitive, and communications skills that humans bring to bear). The efficiency of proficiency scaling through these types of training was seen by some as likely to be domain dependent.

During these discussions, participants highlighted several common research areas necessary to advance IH-MC, such as communication, flexibility and resilience, human-machine models, user experience and system design, testbeds, and data overload.

Communication

For many participants, effective communication represents a significant barrier to advances in human-machine collaboration. According to Kruijff, it would be useful if robots could better explain to humans what they can or cannot do and what they actually do. An inability to effectively communicate this, he added, makes it difficult for humans and machines to gain common ground. In addition to verbal communication, Sidner noted the challenges posed by nonverbal behavior—for example, how might a robot "notice" what a human notices?

Other participants commented on the benefit of further research in how humans and machines communicate and understand intent. Bradshaw emphasized that by improving machine observability or "apparency," humans will be better able to understand a device's intent. In contrast, Veloso observed that cases may exist in which a human does not necessarily need to monitor or understand what the robot is doing, as long as he or she trusts the robot to proactively ask for help when necessary. Oron-Gilad cautioned that effectively conveying intent between two human operators, let alone between humans and robots, is still a challenge. For example, if a software agent incorrectly "guesses" a human's intent, it might unnecessarily automate a task—thus leading to dangerous and unintended consequences.

Padgham proposed the development of a "teaming compact" whereby humans and machines mutually communicate their capabilities, goals, and intentions. Perhaps what is required, she said, is one common and simple language that can be used by any system. Also necessary, Matthias Scheutz added, are feedback mechanisms and intelligent and tangible interfaces between humans and agents. Other participants commented that this feedback should be dynamic so that human-machine collaboration can change over time—for example, as a result of training or changes in familiarity or trust.

This prompted a discussion on whether new ways for robots to communicate with one another could reduce the number of humans in human-robot teams. In response, one participant suggested that challenges of effective robot-to-robot communication would be made simpler by removing the cultural baggage of human communication.

Flexibility and Resilience

Some participants commented that improved human-machine collaboration will require improved flexibility and resilience. For example, Hoffman observed that human-machine interfaces would be improved by engineering for resilience, that is, designed for unanticipated tasks. Such flexibility is particularly important, Jean Scholtz added, as the tasks people do today will not be identical to those being done tomorrow or in five or ten years. Humans can rapidly adapt and apply their capabilities to new situations, so how can this flexibility and learning be applied to robots without significant programming?

Frank Dignum suggested that lessons may be learned from human adaptability—for example, the adaptation of human language to widespread adoption of text messaging. Rather than wait for convergence in, for example, natural language between humans and robots, he proposed a deeper examination into situations in which humans but not robots are able to adapt.

This resilience, Neerincx noted, will require breakthroughs in context-driven adaptive autonomy. Both Hoffman and Sidner commented that such high levels of complex autonomy would first depend on significant breakthroughs in commonsense knowledge and practical manipulation tasks.

Modeling

Another common challenge discussed was the potential benefit of improved human, machine, and shared human-machine models. Goodrich spoke to the difficulties of developing such shared models by describing human and machine dynamic asymmetries in experience, understanding, goals, and capabilities.

Although some participants emphasized the need to provide robots with better models of humans, Scheutz noted the challenges of building correct models of robots for humans. Human models of robots, he said, need to be compatible with the ways humans will interact with them. For example, if a robot does not have good natural language or good visual sensing capabilities, perhaps anthropomorphized robot mouths or eyes will mislead humans to overestimate the robot's capabilities. While highly realistic Geminoid robots exist, Holly Yanco added that the "uncanny valley" factor should also be taken into consideration.

According to Dignum, new social reality models may allow machines to do things "with" humans, and not just "for" humans in a limited role. Many of the workshop participants remarked that it would help for humans to develop social understanding and acceptance of such sophisticated machine capabilities in order for this level of collaboration to occur.

Testbeds and Fielded Systems

Several participants suggested that more and improved large-scale dynamic testbeds (as well as their ongoing evaluation) would benefit many of the previously discussed research issues. Going beyond testbeds, Kruijff observed that some of the challenges the group discussed would best be studied using deployed or fielded systems. For example, some challenges, such as philosophical linguistics issues, are more likely to arise in the field as opposed to the laboratory. True collaboration under stressful circumstances, he noted, cannot be replicated in the laboratory.

"Big Data"

Challenges posed by "data overload" were also highlighted, in the context of improving both human-machine interaction and teamwork, as well as the value of using "big data" to solve large-scale problems.

Some suggested that new ways for teams to share concrete and dynamic information about their environment could provide novel perspectives that lead to interesting and new solutions. For Satoshi Tadokoro, this is particularly relevant in the context of supporting teams composed of one human and multiple robots. This type of coordination, he observed, requires significant amounts of and access to data. In the rescue domain, this means data about human-robot coordination and interaction, as well as the environment.

Morison proposed that significant opportunities exist for breakthroughs in the ways that large-scale robot/sensor data are used to expand the ways humans perceive the world. Systems could be designed, he said, for effective exploration so that relevant information can be quickly extracted.

Shared Resources for Shared Problems

Using IH-MC to solve highly complex problems, Hoffman noted, requires big research budgets, often in harsh economic climates. One path forward, he proposed, might be to choose a single problem large enough to require international funding efforts.

Padgham acknowledged that many fields associated with human-machine collaboration have not been as successful as others in disciplining themselves to combine resources in pursuit of solving larger-scale challenges. In part, this is because long-term funding to support such initiatives has not been as available in this field as it has been in others. As an example of a successful initiative, Sidner posited that the success of the physics research communities in effectively combining resources has, to some degree, been a result of 400 years of maturation within a set of unified fields. Perhaps, Padgham suggested, IHMC efforts to manage a small disaster would be appropriate for international funding and combined large-scale research efforts.

5

Global and Transnational Issues

There are significant real-world problems, Bradshaw proposed, that by their very nature are international in scope and would benefit by participation from researchers from different countries and different disciplines. Sonenberg added that for many large-scale (and potentially international or global) problems, great opportunities exist for collaboration and coordination to meet shared goals. Today, researchers are tapping into the potential to exploit the Web to collect, integrate, and share data in useful ways to support the flow of data from information to knowledge. In addition, new technological capabilities, such as large-scale and massively distributed sensor systems, are allowing researchers to explore new, and potentially global, scales where the "field" has become the "laboratory."

She also referred to the scenario discussions on cross-cultural issues that addressed differing norms regarding "personal space," gender roles and preferences, safety and trust in automation, and communication. Addressing these cultural differences, she noted, would benefit from national and local—as well as global–expertise.

From an international manufacturing and assembly collaboration perspective, Don Mottaz described the challenges of translating process information into other languages and cultures. While current efforts focus on teaching humans, he proposed that machines may one day be used to teach humans from a variety of different cultural backgrounds. Thus spoken, written, tactile, and other teaching strategies will help to incorporate cross-cultural human-machine interaction requirements.

Lakmal Seneviratne discussed the increasing global use of automated tools in surgical environments, as well as long-distance teleoperated robotic surgeries. In addition to time zone differences, technological challenges from computer-robot delays (above 1/100 of a second), operating room team dynamics and hierarchies, and social acceptance of robot surgery tools across cultures by both medical practitioners and patients still present significant obstacles.

In addition to robotic surgery applications, Wagner proposed that medical doctors might beam into rural or underserved hospitals and clinics to conduct physical exams and deploy further specialization. He would like to see these tasks move beyond "skype-on-wheels." Oron-Gilad suggested that environmental context is also an important factor in remote presence. For example, remote participants may not realize that they have beamed into a stressful, unpredictable, or dangerous environment, such as a war zone, thus underappreciating or underutilizing the context in which the local staff is operating.

Ramchurn identified energy management as a global issue in which agents and machines will play a role. As nations shift their focus to renewable energy sources, he suggested, intermittent sources and supply/demand constraints may require agents to have some control of devices (e.g., washing machines) to influence energy usage patterns. In these circumstances, humans would actually be adapting their behavior to agents.

For search and rescue missions, Kruijff commented that cultural considerations come into play when local, national, and regional agencies or organizations need to deliver, share, and coordinate information. For this reason, Tadokoro emphasized multiculturally sensitive data-gathering and -sharing strategies.

Lastly, Sonenberg commented that effectively addressing the social and cultural implications of human-machine collaboration will call for social scientists and anthropologists to work together with engineers. Further, many of these kinds of collaborations will be studied more effectively in natural settings than in the lab.

6

Revisiting the Scenarios

On day three of the workshop, participants were placed into small groups and given the opportunity to revisit previous scenarios for a second round of analysis. Loosely following the DARPA Grand Challenge competition, each group developed a 10-year research proposal on a topic of their own choosing, using two of the earlier scenario discussions as a starting point: Hospital Service Robotics and Preparing For and Managing a Major Disaster. In addition to receiving an unlimited research budget, each group was obligated to rely on the actual expertise of its members. For example, if a group did not possess a natural language expert, its delivered system could not employ sophisticated or innovative natural language.

Group 1: Disaster Management System for a Collapsed Urban Hotel
Moderator: Alex Morison
Group Members: Paul Maglio, Alex Morison, Don Mottaz, Gopal Ramchurn

The moderator, Alex Morison, spoke on behalf of the group. Based on the large-scale volcanic eruption scenario, he discussed the group's development of a Disaster Management System for search and rescue efforts following the collapse of an urban hotel. As a result of the collapse, people are believed to be trapped in the rubble within contained cavities that are not navigable by humans or dogs. The group's system would make effective use of robots to map cavities within the rubble (for size, location, interconnectedness) and coordinate the exploration. Key technological challenges include: mobility, structural stability, communications, environmental awareness, multi-robot coordination, and "big data" sense making. The system's design considerations would very likely include both staged rescue scenarios and real-world rescue efforts with actual rescue personnel.

To provide the mobility necessary for such a system, the group proposed the design of a crawling robot composed of multiple modular sensor units. This "slug-like" robot would consist of a series of sensor arrays; for example, one module might be an antenna system to improve communication capabilities.

In addition to navigating confined and unstable spaces, multiple robots—each with different perspectives—would provide improved spatial awareness; new reasoning functions could provide 3-D mapping capabilities.

Morison noted that the system's success will depend on how well multiple robots can work together as a team. In fact, the group identified multi-robot teamwork as the group's most significant challenge, citing the current lack of breakthroughs in communications protocols and multi-agent coordination. Effective multi-robot coordination becomes especially critical in post-disaster environments that are often resource limited and unpredictable. Under some circumstances, humans would assume a larger or primary role in coordination efforts—for example, under system failure or when human expertise is required. In cases where humans and robots share responsibilities, automated reasoning would be combined with human reasoning.

Lastly, the group observed that as robots develop increased autonomous capabilities, there may be a push for increased autonomous decision making. The group questioned what if anything might limit such autonomy. For example, what ethical considerations exist for human robot rescue teams (with varying various levels of autonomous capabilities) that triage lost or injured individuals?

Group 2: Team Clean
Moderator: Michael Beetz
Group Member: Michael Beetz, Andreas Hofmann, Mark Neerincx, Liz Sonenberg

Michael Beetz, the moderator, provided a summary of the group's discussions. Beetz indicated that the group focused its efforts on designing a home robotic cleaning team, "Team Clean," composed of multiple machines (e.g., humanoid robot, vacuum cleaner, small UAV to "map" the environment), and potentially a human director. The team would be capable of accomplishing a number of tasks with varying degrees of difficulty, from cleaning bathrooms to washing dishes, vacuuming, and doing the laundry.

To do this, a number of research challenges would be addressed, including: practical task manipulation (e.g., picking up fragile objects), smooth locomotion and navigation in a dynamic environment (e.g., going up stairs and opening doors), safety (e.g., not getting in the way of residents or pets), human-robot communication, and social robotics. In addition, machines would have to be able to learn and recover from mistakes and possess sufficient knowledge intensiveness (e.g., to go from an abstract task "to clean up" to understanding how clean is "clean enough").

Some tasks, Beetz acknowledged, would require varying degrees of interaction between machines and residents. In some cases, a robot may request feedback from the resident. For example, a robot might ask whether a dirty glass situated near the resident is currently being used or in need of washing. In other cases, the system should be adaptable if it is re-tasked by the resident. This could occur if the resident's cleaning expectations differ from those of the robot. The robot may also need to resolve conflicting resident demands—for example, balancing a parental request to "pick things up off the floor" and a teenager's request to "leave the bedroom as it is." To deal with this situation, the group proposed a system with one "chief," as well as an organizational structure to deal with conflicting goals.

Lastly, the group identified performance evaluation as a significant element of the system. For example, how many tasks were accomplished and in what time frame? How well did the team function? As the team evolved, the system would be scalable to accomplish a wider range of tasks.

Group 3: Biped Hospital Companion Robot
Moderator: Candy Sidner
Group Members: Robert Hoffman, Lakmal Seneviratne, Candy Sidner, Rong Xiong

The moderator, Candy Sidner, provided a description of the group's proposal for creating a biped hospital companion robot (based on an earlier discussion of hospital service robotics). This robot would undertake personal care activities (e.g., dressing and bathing patients and picking up laundry) and provide mobility/balance support by preventing mobility-related accidents and catching patients who are falling. In addition to physical manipulation requirements, some basis for human-robot communication is required and humans need to be comfortable receiving robotic assistance. For this reason, human-robot trust is an important systems requirement.

To accomplish these tasks, the proposed biped robot would be designed with articulated, touch-sensitive hands and somewhat soft bodies with suitable, nonaversive "skins." In addition, visual recognition would be integrated with touch and task-manipulation capabilities. Communication between the robot and patient would be computer-controlled and employ simple dialogue—for example, questions that can be answered with a "yes" or "no" or with a very short statement.

Sidner added that algorithms, such as those used to predict the movements of rapidly traveling Ping-Pong balls, would be tuned and applied to predict when a human is falling and to respond appropriately. This would require the robot to distinguish not only between types of falling (e.g., falling while conscious or unconscious), but also between similar actions (e.g., falling versus bending over to pick something up). For some frail individuals, the group noted, the line between falling and bending over is thin. By merging robot companion and robot assistant technologies, the robot could also act as an instructor or coach for patients. For example, a robot that has learned to balance itself could not only help feeble patients cross hospital floors but could also act as a physical therapy coach.

As a part of its design, the group would also develop a number of testbeds to assess both communication and trust issues, as well as appropriate and safe interactions between robots and patients. Questions addressed would include: How does the nature of human-robot communication and interaction change when robots are working with patients who may be sick, feeble, or physically or cognitively impaired? How anthropomorphic should a robot companion be, and should it be more or less anthropomorphic if it is engaging in a conversation or dressing/undressing a patient? How might robot companions work with other robot companions in this environment?

Group 4: The Robotic Patient Advocate
Moderator: Michael Freed
Group Members: Michael Freed, Yukie Nagai, Jean Scholtz, Satoshi Tadokoro, Manuela Veloso

The moderator, Michael Freed, spoke on behalf of the group. Using the medical service robots as a starting point, Freed described the group's proposal for creating a robotic patient advocate that would work either as an intermediary between the patient and hospital staff or directly with patients. The robotic patient advocate would keep nurses up-to-date (e.g.., monitor and report changes in patient physical or emotional states), provide continuity when nurses change shifts or when patients are assigned new doctors, and communicate with nurses when patients are asleep or unable to effectively communicate. In addition, the advocate would directly provide information to confused or forgetful patients (e.g., asking "Why am I being wheeled to Room 108?" or "Have I taken my medication already?"). The advocate would also support medical staff when the patient required encouragement. Lastly, the advocate would run interference with visitors and people who stay too long or get in the way of medical staff.

As Freed explained, the advocate would leverage the group's collective experience in autonomous systems, communications and dialogue, human emotion, machine-human interaction, and performance evaluation of both robots and humans. Although some of these capabilities were possible using conventional technologies, six key breakthroughs would be required. (1) *Dialog*: The advocate should be capable of high-level discussions with people possessing different knowledge, motives, and cultures. (2) *Multimodal Sensing*: The advocate should be able to tap into—via many and complex sensors—a hospital's data-rich environments to access a patient's medical records, real-time physiological conditions, test results, and schedules. (3) *Strategic Planning*: The advocate should take action by balancing a patient's immediate goals and requests with long-term patient support that considers legal and safety issues. (4) *Safe Navigation*: The advocate should navigate a complex environment of constantly changing people, carts, beds, and equipment. (5) *Social Understanding*: This might require the advocate to know when and how to "shoo" away visitors who are unwanted or have stayed too long. (6) *Multi-Persona Negotiation*: The advocate should be a middleman or "middlerobot" among a floating team consisting of patients, family members, doctors, nurses, and others.

The group acknowledged that evaluation of the system was critical; thus, the advocate would be developed first with limited capabilities that would be expanded on the basis of experience and learning.

Group 5: Providing Post-Disaster Basic Services
Moderator: Lin Padgham
Group Members: Tal Oron-Gilad, Lin Padgham, Dirk Schulz, Holly Yanco

Lin Padgham, the moderator, provided the description of the group's proposal. As one component of the volcanic eruption disaster-management scenario, Padgham described the group's design of an information management system to provide basic services, such as communications, food, power, and water, in the first week following a major disaster. The system would not provide total coordination across the entire disaster management value chain, but rather would provide on-the-ground individuals with decision support.

Such a decision-support system would require data inputs from numerous sources, including cell phones, sensors, weather reports, and UAVs. The system would take in data in a variety of formats and then organize and share those data with a range of specialized users. For example, data inputs from UAVs that show downed power lines could be used to coordinate prompt robot deliveries of electrical and other power sources to neighborhoods lacking electricity. Effectively distributing and acting on this information will require simple yet specialized human-machine interfaces.

Ongoing access to massive amounts of parallel data would allow management officials to better prioritize their attention and efforts—for example, whether to immediately evacuate a neighborhood or to first restore basic infrastructure. Padgham added that the system could be used as a simulation tool in advance of a disaster to improve emergency management response. By assessing the efficacy of different communications protocols and of evacuation routes under different environmental and social circumstances, authorities can identify where critical post-disaster response failures are likely to occur.

The system would also make "individualized" information available to both specialized users (e.g., UAV operators with specific data needs to survey for downed power lines) and to untrained users who are stranded in their homes with limited food and water. Although acknowledging that such a system would provide complex decision support, the group noted that human judgment will always remain key.

A

Workshop Participants

MICHAEL BEETZ, Technische Universitat Muenchen
JEFFREY M. BRADSHAW, Florida Institute for Human and Machine Cognition
FRANK DIGNUM, Utrecht University
TERRY FONG, NASA Ames
MICHAEL FREED, SRI International
TAL ORON-GILAD, Ben-Gurion University
MICHAEL GOODRICH, Brigham Young University
ROBERT HOFFMAN, Florida Institute for Human and Machine Cognition
ANDREAS HOFMANN, Vecna Technologies
GEERT-JAN (GJ) KRUIJFF, Deutsches Forschungszentrum für Künstliche Intelligenz
PAUL MAGLIO, IBM Research and University of California, Merced
ALEXANDER MORISON, The Ohio State University
DON MOTTAZ, The Boeing Company
YUKIE NAGAI, Osaka University
DANIELE NARDI, University of Rome
MARK NEERINCX, Delft University of Technology
LIN PADGHAM, Royal Melbourne Institute of Technology
SARVAPALI (GOPAL) RAMCHURN, Southampton University
MATTHIAS SCHEUTZ, Tufts University
JEAN SCHOLTZ, Pacific Northwest National Laboratory
DIRK SHULZ, Fraunhofer Institute for Communications, Information Processing and Ergonomics
LAKMAL SENEVIRATNE, Khalifa University and King's College London
CANDY SIDNER, Worcester Polytechnic University
LIZ SONENBERG, University of Melbourne
SATOSHI TADOKORO, Tohoku University

MANUELA VELOSO, Carnegie Mellon University
RONG XIONG, Zhejiang University
TOM WAGNER, iRobot
BRIAN WILLIAMS, Massachusetts Institute of Technology
HOLLY YANCO, University of Massachusetts, Lowell

B
Workshop Agenda

Monday, June 11th

6:00 – 7:30 PM	Welcome reception (State Plaza Hotel, Ambassador Room)

Tuesday, June 12th

8:00 – 8:30 AM	Breakfast (Room 120)
8:30 – 8:45 AM	Welcome and setting the stage
8:45 – 9:45 AM	Participant introductions
9:45 – 10:00 AM	Introduction to scenario exercise Moderator: Brian Williams
10:00 – 12:30 PM	Breakout Groups: Real-World Applications of Intelligent Human-Machine Collaboration (IH-MC) Scenario A: Disaster Management Moderator: Michael Goodrich Scenario B: Small-Lot Agile Manufacturing Moderator: Matthias Scheutz Scenario C: Hospital Service Robotics Moderator: Candy Sidner Scenario D: Virtual Team Training Moderator: Mark Neerincx Scenario E: Personal Satellite Assistants Moderator: Terry Fong
12:30 – 1:30 PM	Lunch

1:30 – 3:00 PM	Group discussion (Moderator: Jean Scholtz)
	Breakout groups A, B, and C report back on findings from earlier scenario exercise (30 minutes each)
3:00 – 3:15 PM	Break (refreshments available)
3:15 – 4:15 PM	Group discussion (Moderator: Tal Oron-Gilad)
	Breakout groups D and E report back on findings from earlier scenario exercise (30 minutes each)
4:15 – 4:30 PM	Break
4:30 – 5:30 PM	Group discussion (Moderator: Lin Padgham) • What international, global, or cross-cultural considerations were raised during your scenario discussions? • What are the benefits of intelligent human-machine collaboration vs. traditional autonomy? • What are some of the commonalities in human-machine issues that were raised across the scenarios? • What are the issues that were not raised? • What are the biggest overall research challenges? Which of these challenges would require significant breakthroughs? Which of these breakthroughs are unlikely to occur in the next ten years? In twenty years? Based on this discussion, workshop participants will select and self-organize into 5 topics for the next day's Collaboration Panels.
5:30 PM	End of day one.

Wednesday, June 13th

8:00 – 8:30 AM	Breakfast
8:30 – 10:400 AM	Reflections from day one
8:45 – 9:45 AM	Panel Breakout Groups

APPENDIX B

	Participants will meet with their respective panels (organized the previous afternoon) to organize a 30-minute discussion. Each panel should create a PowerPoint presentation for the discussion.
10:00 – 10:30 AM	Panel I: (30 minutes)
10:30 – 10:45 AM	Break (refreshments available)
10:45 – 11:15 AM	Panel II: (30 minutes)
11:15 – 11:45 AM	Panel III: (30 minutes)
11:45 – 12:15 PM	Panel IV: (30 minutes)
12:15 – 1:30 PM	Lunch (Group picture at Albert Einstein statue)
1:30 – 2:30 PM	Group discussion (Moderator: Manuela Veloso)
	What kinds of breakthroughs would be game changers for significantly improved intelligent human-machine collaboration? What are the implications of these breakthroughs for national and global security, competitiveness, and human well-being?
2:30 – 3:00 PM	Break (refreshments available)
3:00 – 4:00 PM	Group discussion (Moderator: Liz Sonenberg)
	What are the global (or transnational) challenges that intelligent human-machine collaboration can help to solve?
4:00 – 5:00 PM	Group discussion (Moderator: GJ Kruijff)
	Summary of challenges and solutions discussed throughout workshop
5:00 PM	End of day two.

Thursday, June 14th

8:00 – 8:30 AM	Breakfast

8:30 – 10:30 AM	Research Topics in Intelligent Human-Machine Collaboration (IH-MC)
	Session 1: Sociocognitive Issues Moderator: Jeff Bradshaw
	Yukie Nagai, Osaka University *Robots That Learn to Communicate with Humans*
	Alex Morison, Ohio State University *Expanding Human Perception and Attention to New Spatial-Temporal Scale through Networks of Sensor Systems*
	Candy Sidner, Worcester Polytechnic University *Agents for Long-Term Relationships with Isolated Older Adults*
	Frank Dignum, Utrecht University *Interaction in Context*
10:30 – 10:45 AM	Break (refreshments available)
10:45 – 11:45 PM	Research Topics in IH-MC (continued)
	Session 2: Challenging Applications Moderator: GJ Kruijff
	Lakmal Seneviratne, Khalifa University & King's College London *Force Feedback and Haptic Interfaces during Robot-Assisted Surgical Interventions*
	Rong Xiong, Zhejiang University *A Study on Humanoid Robots Playing Table Tennis*
11:45 – 1:30 PM	Working Lunch (meal tickets at cafeteria)
	Breakouts: Grand Challenges/Scenario Revisits • Describe your scenario and the aspect(s) you will address. • What is your conceptual architecture?

APPENDIX B 37

	• What are your innovative claims?
1:30 -2:45 PM	Report Back
2:45 – 3:00 PM	Break (refreshments available)
3:00 – 4:30 PM	Research Topics in IH-MC (continued)
	Session 3: Learning and Adaptation in Dynamic Settings Moderator: Jeff Bradshaw
	Michael Freed, SRI International *A Virtual Assistant for E-mail Overload*
	Satoshi Tadokoro, Tohoku University *The Disaster Response Robot Named "Quince" and Lessons at the Fukushima-Daiichi Nuclear Power Plant Accident*
	Michael Goodrich, Brigham Young University *Autonomy, Interaction, and Collaboration: A WiSAR Perspective*
4:30 – 5:00 PM	Research Topics in IH-MC (continued)
	Session 4: Human-Machine Interaction and Teaming Moderator: Brian Williams
	Holly Yanco, University of Massachusetts Lowell *Human-in-the-Loop Control of Robot Systems*
5:00 – 5:15 PM	Final discussion
5:15 PM	Meeting adjourned

C

Presentation Abstracts

Session 1: Sociocognitive Issues

Yukie Nagai, Osaka University
Title: Robots that learn to communicate with humans

Abstract: How can robots learn to communicate with humans? How can they acquire the ability to read the intentions of humans? In order to collaborate with human partners, robots need to understand what the goal of the partner's action is. Inspired by studies of developmental psychology and neuroscience, our lab has been developing robots that learn to communicate with others based on the mirror neuron system (MNS). The MNS plays a central role in understanding the goal of the other's actions and imitating them. We have hypothesized that the MNS emerges through sensorimotor learning accompanied by perceptual development; immature perception in the early stages of development enables robots as well as infants to find the correspondence between the self and other (an important property of the MNS). My talk will present the results of the robotics experiment to verify this hypothesis and also the results of an additional experiment, which analyzes the microscopic structure of caregiver-infant interaction in order to better understand the developmental mechanism of infants. In this paper, I emphasize the importance of perceptual and motor immaturity in leading to further- and better-organized cognitive development.

Alex Morison, Ohio State University
Title: Expanding human perception and attention to new spatial-temporal scales through networks of sensor systems

Abstract: Ubiquitous sensing capabilities create the potential to expand human reach to new spatial-temporal scales, but to date the potential is unrealized. Models of how human perceptual systems function successfully to manage multiple data streams and directly apprehend the world have inspired new technologies and visualizations to overcome data overload and release the power of new human-sensor systems.

Candy Sidner, Worcester Polytechnic University
Title: Agents for long-term relationships with isolated older adults

Abstract: We are exploring the development of virtual agents who "live" in the homes of socially isolated older adults for extended periods of time. Our agent reasons about activities that are appropriate to undertake with the adult as its relationship changes, from stranger to something one might call "companion" in the course of daily interactions. In this talk, I will discuss the relationship manager that reasons about the relationship and plans activities, and the real-time collaboration manager, which puts those plans into effect while also reasoning about time and the time available to complete those plans. I will also discuss experiments with older adults in their homes, who use prototype agents to help us discover what the agent can best be doing with adults.

Frank Dignum, Utrecht University
Title: Interaction in context

Abstract: When people interact they use context to both express and interpret the meaning of the information they want to exchange. Unfortunately, there are many overlapping contexts that might be active at the same time. Thus, choosing the right context to generate or interpret a message is a complex but very important issue for human-machine collaboration, especially for human, agent, and robot teams.

Session 2: Challenging Applications

Lakmal Seneviratne, Khalifa University, Abu Dhabi, UAE, and King's College London, UK
Title: Force feedback and haptic interfaces during robot-assisted surgical interventions

Abstract: In recent years there have been significant advances in robot-assisted minimally invasive surgical (MIS) procedures. However, although robot-assisted MIS represents significant improvements over traditional MIS, it does not provide the surgeon with a sense of touch from the operating interface. Many robotic surgical applications require active interactions with complex dynamic environments such as soft tissue. A fundamental understanding of the interaction dynamics between the surgical system and the environment is an essential element in intelligent surgeon-robot collaboration. The sensing of forces at the robot-tissue interface is a very challenging research problem. In this presentation we survey a number of force and stiffness sensors developed for surgical robotic systems. These include force and stiffness sensors based on fiber-optic and pneumatic technologies. We explore finite element (FE) modeling of the robot-tissue interface, including inverse FE models for identifying tissue properties for diagnosis. The use of haptic interfaces at the surgeon-master interface is also investigated.

Rong Xiong, Zhejiang University, China
Title: A study on humanoid robots playing table tennis

Abstract: Over the past twenty years, the research on humanoid robots has rapidly advanced, and various humanoid robots have been developed. They can walk, run, dance, play Taiji, etc. The ongoing research on humanoids is moving toward complex task performing in different environments, such as providing domestic service in a home environment or collaborating with human beings to move heavy objects. We take table tennis playing as an entry point to explore related technologies, because both intelligent interaction and dynamic response, which are fundamental factors for future service robots, are required but challenging issues in such a task. We have proposed algorithms for fast visual recognition and accurate trajectory prediction of a Ping-Pong ball and coordinative motion planning and balance maintenance of the humanoid robot, and we have developed a real-time field bus to meet the requirements for quick response. Now the two 165 cm-tall humanoid robots we developed, "Wu" and "Kong," can play table tennis continuously with each other and with amateur human players. This research topic also provides an interesting point of view for studies on autonomous cooperative or competitive interaction between robots or between a human and a robot. For example, how should the robot learn play motions and play strategies from human players? How should the robot vary its play motion and strategies depending on its real-time perception?

Session 3: Learning and Adaptation in Dynamic Settings

Michael Freed, SRI International
Title: A virtual assistant for e-mail overload

Abstract: E-mail client software is widely used for personal task management, a purpose for which it was not designed and is poorly suited. Past attempts to remedy the problem have focused on adding task management features to the client user interaction. RADAR uses an alternative approach modeled on a trusted human assistant who reads mail, identifies task-relevant message content, and helps manage and execute tasks. This talk describes the integration of diverse AI technologies and presents results from human evaluation studies comparing RADAR user performance to unaided commercial-off-the-shelf tool users and users partnered with a human assistant. As machine learning plays a central role in many system components, we also compare versions of RADAR with and without learning. Our tests show a clear advantage for learning-enabled RADAR over all other test conditions.

Satoshi Tadokoro, Tohoku University
Title: The disaster response robot named "Quince" and lessons at the Fukushima-Daiichi nuclear power plant accident

Abstract: The accident at the Fukushima-Daiichi power plant, caused by the tsunami on March 11, 2011, resulted in a meltdown of nuclear fuel and in the hydrogen explosion of nuclear reactor buildings. Several robotic systems were applied to stabilize the situation there. A disaster response robot, Quince, which was developed by the presenter's group, was utilized for surveillance of the 2nd through 5th floors of the nuclear reactor buildings and achieved a certain contribution to their cool shutdown. It was a typical human-machine collaboration task. Both the researcher and engineer side and the user side learned many things in order to apply the robotic system to the unknown environment. This talk introduces an overview of this mission and lessons learned.

Michael Goodrich, Brigham Young University
Title: Autonomy, interaction, and collaboration: A WiSAR perspective

Abstract: Based on discussions at the workshop, an operational definition of "collaboration" was created. Collaboration is a multi-agent problem that emerges when agents have asymmetric information, asymmetric goals, and asymmetric capabilities. These asymmetries enable agents to share resources to solve a problem that the agents couldn't solve independently, but these asymmetries also lead to potential conflicts of interest or points of confusion. This definition of collaboration sheds light on how a technical search team can use an unmanned aerial vehicle to support wilderness search and rescue. Technologies developed to support wilderness search and rescue teams can benefit by supporting the collaborative nature of the team. Importantly, collaboration can be seen as the (re)unification of two threads of research that were both present in Sheridan and Verplank's classic report, which is known for defining levels of autonomy but split the discussion into research on these levels and research on interaction design.

Session 4: Human-Machine Interaction and Teaming

Holly Yanco, University of Massachusetts Lowell
Title: Human-in-the-loop control of robot systems

Abstract: Robots navigating in difficult and dynamic environments often need assistance from human operators or supervisors, either in the form of teleoperation or occasional interventions when the robot cannot handle the current situation autonomously. Even in office environments, robots may need to ask for directions in unknown buildings. In this presentation, I will discuss my lab's research on the best practices for controlling both individual robots and groups of robots, in applications ranging from assistive technology to telepresence to search and rescue. A number of methods for this type of human-robot interaction (HRI), including large and small multi-touch devices, software-based operator control units (softOCUs), haptics, and natural language, will be presented. I will also discuss how we can improve HRI by modeling a user's current level of trust in a robot system.